Exploring The BUILDING BLOCKS of Science

Book K

ACTIVITY BOOK

REBECCA W. KELLER, PhD

Illustrations: Janet Moneymaker

Exploring the Building Blocks of Science Book K Activity Book
ISBN 978-1-941181-25-6

Published by Gravitas Publications Inc.
www.realscience4kids.com
www.gravitaspublications.com

CHEMISTRY

The study of the stuff we can see, touch, taste, smell, and hear

Meet members of The Atom Family

Carbon

Nitrogen

Hydrogen

1. color Carbon dark purple or black

2. color Nitrogen blue

3. color Hydrogen gray

What is made of atoms?

Tigers are made of atoms

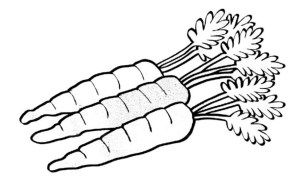

Carrots are made of atoms

Jellyfish are made of atoms

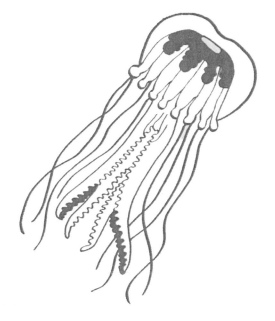

More members of The Atom Family

Phosphorus

Boron

Neon

1. color Phosphorus orange

2. color Boron light pink

3. color Neon light blue

What is made of atoms?

Starfish are made of atoms

Turtles are made of atoms

Cars and ice cream and people
are made of atoms

More members of The Atom Family

Oxygen

Sodium

Silicon

1. color Oxygen red

2. color Sodium dark blue

3. color Silicon orange

Everything you can touch, see, taste, smell, and hear is made of atoms.

Draw something you can taste

More members of The Atom Family

Chlorine

Magnesium

Lithium

1. color Chlorine green

2. color Magnesium dark green

3. color Lithium light purple

What is made of atoms?

Cereal is made of atoms

Birds are made of atoms

Rocks are made of atoms

Everything you can touch, see, taste, smell, and hear is made of atoms.

Draw something you can see

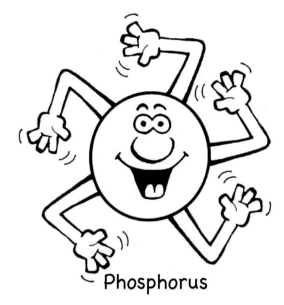

Phosphorus

How many arms does Phosphorus have?

_____

Color Phosphorus yellow

Neon

How many arms does Neon have?

_____

Color Neon light blue

Sodium

How many arms does Sodium have?

_____

Color Sodium dark blue

How many arms does Silicon have?

_____

Color Silicon orange

Silicon

How many arms does Magnesium have?

_____

Color Magnesium dark green

Magnesium

How many arms does Lithium have?

_____

Color Lithium light purple

Lithium

Everything you can touch, see, taste, smell, and hear is made of atoms.

Draw something you can smell

Atoms link arms to make everything we can see, touch, taste, and smell.

Sodium and Chlorine link arms to make
Table Salt

Sodium

Chlorine

Table Salt

1. color Sodium light blue

2. color Chlorine green

3. color Table Salt light blue and green

Everything you can touch, see, taste, smell, and hear is made of atoms.

Draw something you can taste

Everything you can touch, see, taste, smell, and hear is made of atoms.

Draw something you can smell

What is made of atoms?

(fill in the blank and draw a picture)

_____ is made of atoms

_____ is made of atoms

_____ is made of atoms

Which one is not like the other?

1. circle the one that is not like the other

2. color Argon dark blue

3. color Chlorine green

4. color Neon light blue

Everything you can touch, see, taste, smell, and hear is made of atoms.

Draw something you can see

Atoms link arms to make everything we can see, touch, taste, and smell.

Silicon and four Oxygen atoms link arms to make rocks

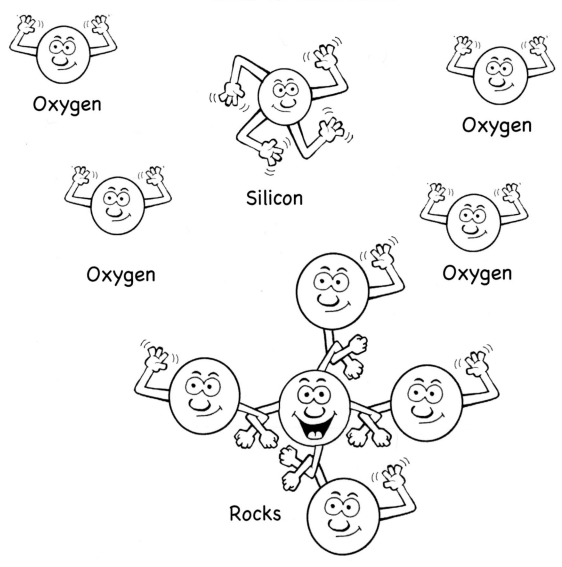

Oxygen

Silicon

Oxygen

Oxygen

Oxygen

Rocks

1. color Silicon orange

2. color all four Oxygen atoms red

3. color Rocks orange and red

Which one is not like the others?

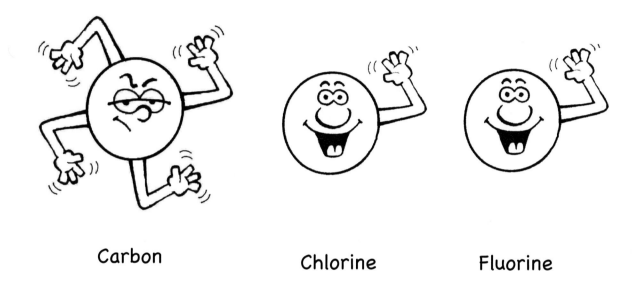

Carbon   Chlorine   Fluorine

1. circle the one that is not like the others

2. color Carbon dark purple or black

3. color Chlorine green

4. color Fluorine light green

What is made of atoms?
(fill in the blank and draw a picture)

_____ is made of atoms

_____ is made of atoms

_____ is made of atoms

Atoms link arms to make everything we can see, touch, taste, and smell.

Carbon, Oxygen and Hydrogen atoms link arms to make sugar

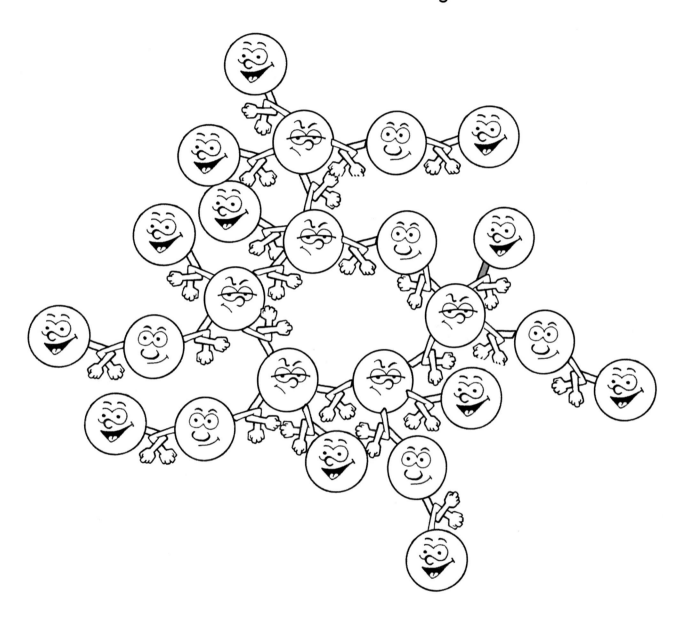

1. Can you find the Carbon atoms? There are 6. Color them dark purple or black.

2. Can you find the Oxygen atoms? There are 6. Color them red.

3. Can you find the Hydrogen atoms? There are 12. Color them gray.

Which two are similar?

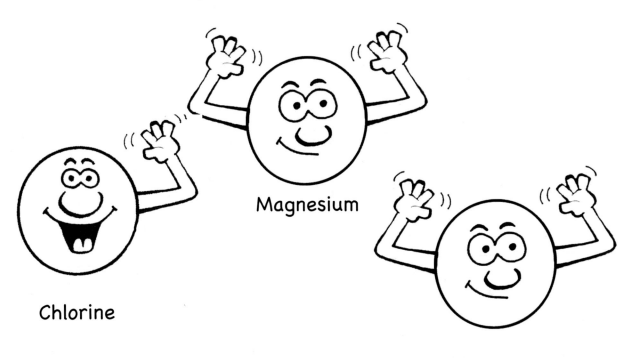

Magnesium

Chlorine

Beryllium

1. circle the two that are similar

2. color Magnesium dark greenk

3. color Chlorine green

4. color Beryllium light green

Everything you can touch, see, taste, smell, and hear is made of atoms.

Draw something you can hear

More members of The Atom Family

Sulfur

Aluminum

Helium

1. color Sulfur yellow

2. color Aluminum pink

3. color Helium sky blue

What is made of atoms?

Bananas are made of atoms

Beetles are made of atoms

Cats are made of atoms

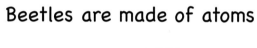

Which two are similar?

Helium

Aluminum

Boron

1. circle the two that are similar

2. color Aluminum pink

3. color Helium sky blue

4. color Boron light pink

What is made of atoms?

(fill in the blank and draw a picture)

_____ is made of atoms

_____ is made of atoms

_____ is made of atoms

How many arms does Aluminum have?

_____

Color Aluminum pink

Aluminum

How many arms does Helium have?

_____

Color Helium sky blue

Helium

How many arms does Sulfur have?

_____

Color Sulfur yellow

Sulfur

Everything you can touch, see, taste, smell, and hear is made of atoms.

Draw something you can touch

Which two are similar?

Sulfur

Argon

Oxygen

1. circle the two that are similar

2. color Sulfur yellow

3. color Argon blue

4. color Oxygen red

Everything you can touch, see, taste, smell, and hear is made of atoms.

Draw something you can smell

Circle the atoms with the same number of arms.

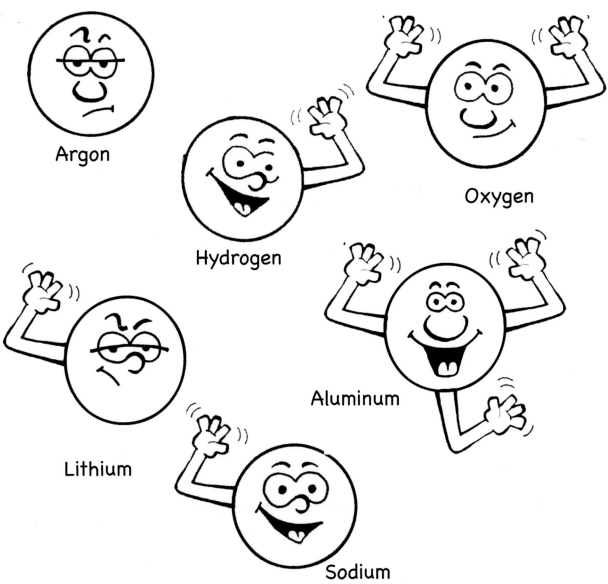

1. color Hydrogen gray

2. color Lithium purple

3. color Argon blue

4. color Oxygen red

5. color Sodium dark blue

6. color Aluminum pink

Where is Argon?

1. find Argon
2. color Argon blue
3. circle Argon
4 color the other atoms

What is made of atoms?

(fill in the blank and draw a picture)

_____ is made of atoms

_____ is made of atoms

_____ is made of atoms

What is water made of?

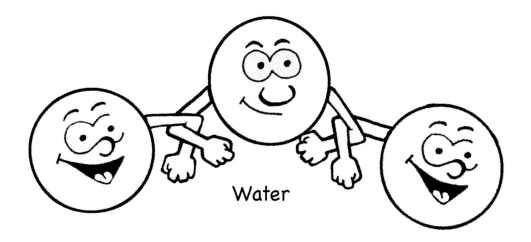

Water

Name the atoms that make water

(hint: There is one of one kind and two of another kind)

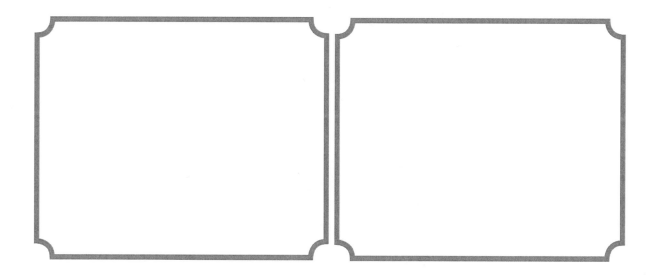

Color the atoms in water

BIOLOGY

The study of

LIFE

What is alive?

Ladybugs are alive

Caterpillars are alive

Ameoebas are alive

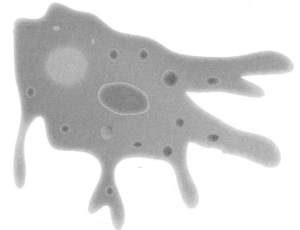

Circle the things that are alive.
Color the pictures

basketball

mama
and baby
kangaroo

turtle

stuffed
bear

lizard

What is alive?
(fill in the blank and draw a picture)

_____ is alive

_____ is alive

_____ is alive

What is alive?

Color the pictures

Rocks are NOT alive

Worms are alive

Alligators are alive

Things that are alive have babies.

Color the picture

Take a Nature Walk

Go into your back yard and draw what you see. Circle
what is alive and put a box around what is not alive.

What is alive?

(fill in the blank and draw a picture)

_____ is alive

_____ is alive

_____ is alive

But a box around the things that are not alive.

Color the pictures

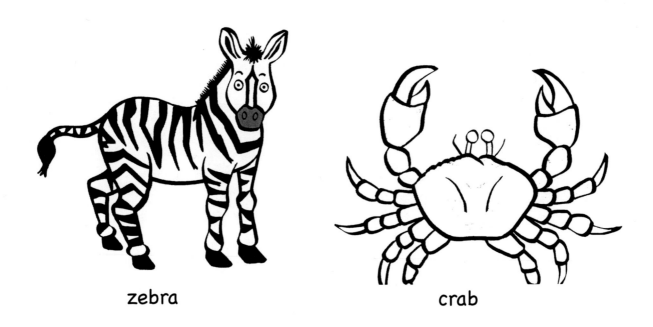

zebra   crab

fire in fireplace

Some living things are large.
Some living things are small.

Color the picture

1. put a box around the smallest living thing in the picture
2. circle the largest living thing in the picture

This is a virus.

It can make you sick with a cold.

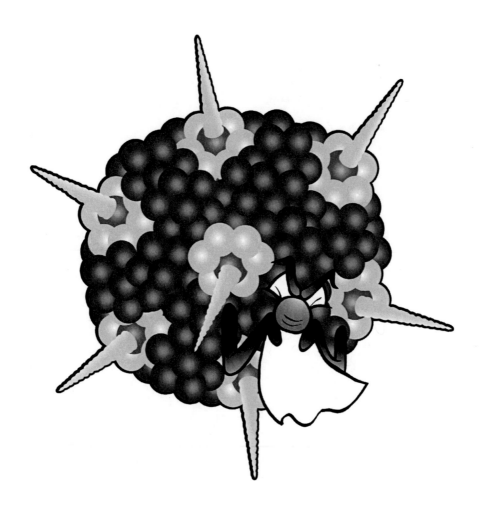

Viruses are very, very small.

Take a Nature Walk

Go to a park and draw what you see. Circle what is alive and put a box around what is not alive.

What is not alive?
(fill in the blank and draw a picture)

_____ is not alive

_____ is not alive

_____ is not alive

These are bacteria.

Some bacteria are good.

Some bacteria can make you sick.

Bacteria are larger than viruses but
are still very small.

The bacteria inside you
are good and keep you
healthy.

A mouse is smaller than a girl.

A baby plant is smaller than a mouse.

A bacteria is smaller than a baby plant.

Some things that are alive can move.

Color the picture

What is alive?
Color the pictures

Toy cars are NOT alive

Bacteria are alive

Tea kettles are NOT alive

Take a Nature Walk

Go to the forest and draw what you see. Circle what is alive and put a box around what is not alive.

Which one of these is
not like the other?

Color the pictures

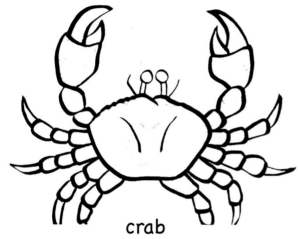

crab

rocks

bacteria

What is alive?
(fill in the blank and draw a picture)

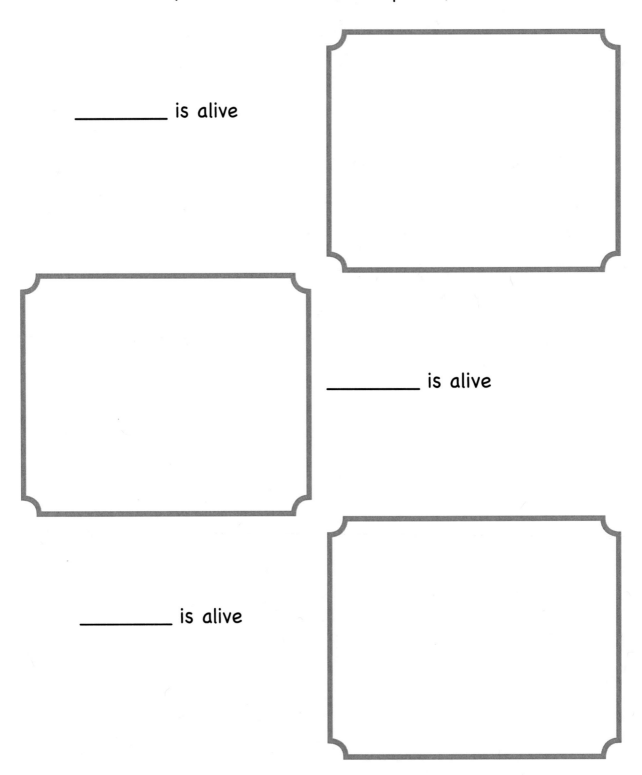

_____ is alive

_____ is alive

_____ is alive

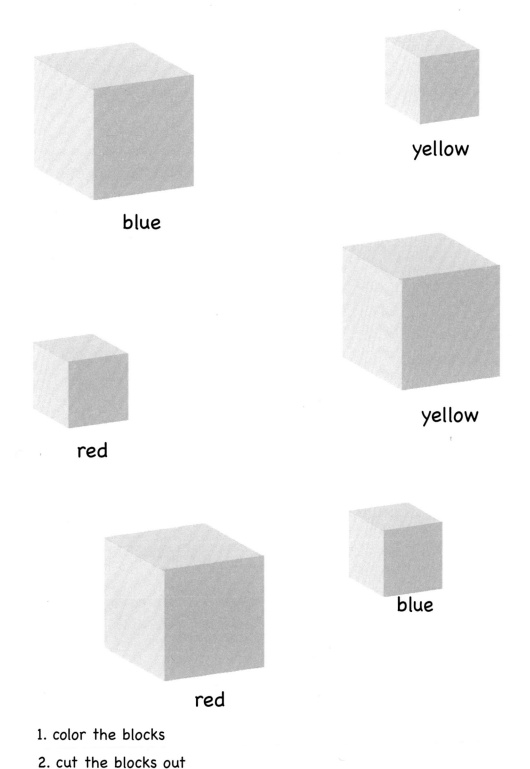

blue

yellow

red

yellow

red

blue

1. color the blocks

2. cut the blocks out

3. use the blocks for the next few activites

Sort the blocks

Put all the large blocks below

Sort the blocks

Put all the blue blocks below

Sort the blocks

Put all the yellow blocks below

Sort the blocks

Put all the red blocks below

Sort the blocks

Put all the small blocks below

How do you sort living things?

1. color the plant green
2. cut the frog green
3. color the rest of the picture

Starfish are grouped with other starfish.

Color the pictures

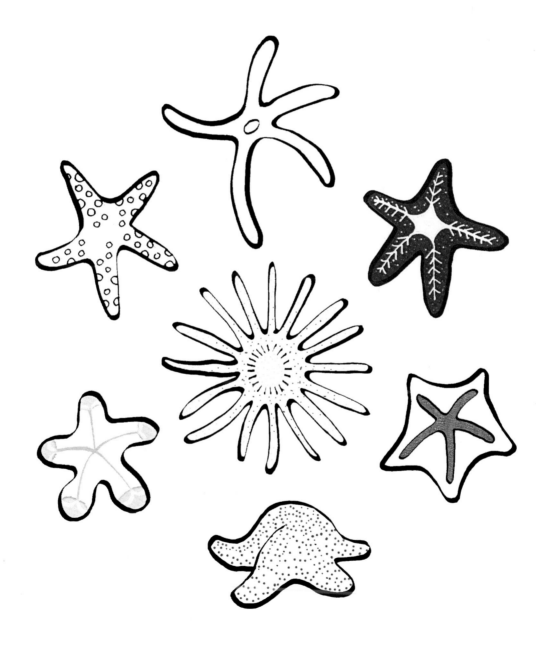

Frogs like to live near water.

Color the picture

Frogs lay eggs that turn into tadpoles.

Color the picture

Tadpoles turn into frogs.

Color the picture

Reptiles are grouped with other reptiles.

Color the pictures

What is alive?

(fill in the blank and draw a picture)

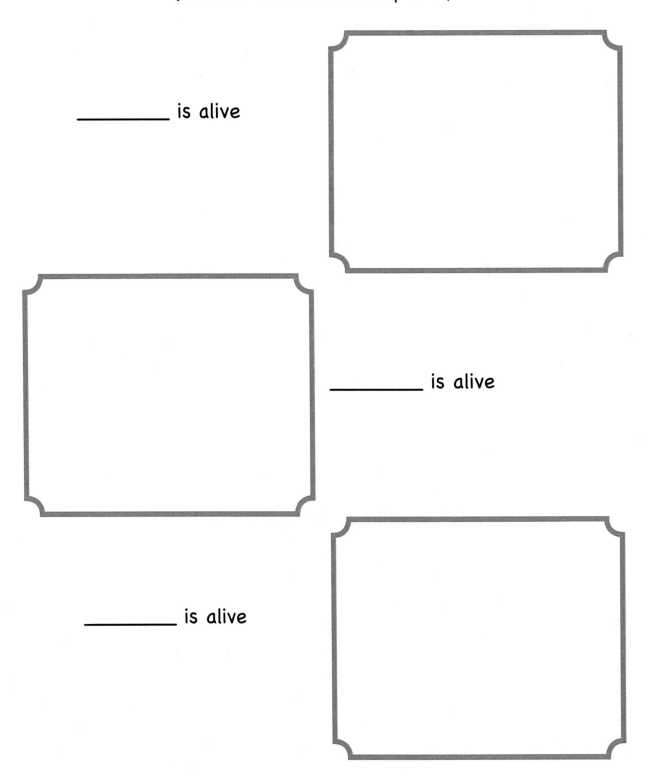

_____ is alive

_____ is alive

_____ is alive

Animals are grouped with other animals.

Color the pictures

How far can you throw a ball?

Color the picture

Go outside and throw a ball as far as you can.

Use your feet to measure how far you threw the ball. Place one foot in front of the other and count your steps as you walk.

How many feet?

What goes up must come down?

Go outside and throw a baseball up in the air.

Does it come down? Draw what you discovered.

Roll a small ball

Color the picture

Is it easy or hard to roll a small ball?

Roll a big ball
Color the picture

Was rolling a big ball easier or harder than a rolling a small ball?

Up the hill
Color the picture

Draw a big ball in the wagon. Is it hard to pull a big ball up the hill?

What goes up must come down?

Go outside and throw a ping pong ball up in the air.

Does it come down? Draw what you discovered.

Up the hill

Color the picture

Draw a small ball in the wagon. Is it hard to pull a small ball up the hill?

How far can you shoot a skinny rubber band?

Color the picture

How many feet?

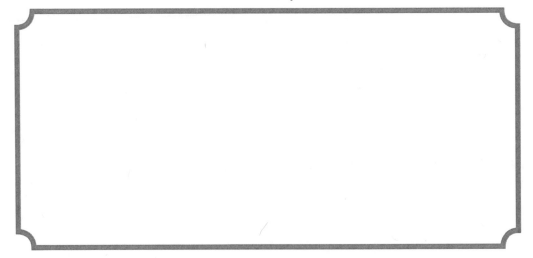

Can you squeeze a marshmallow?

Color the picture

Circle YES or NO

What goes up must come down?

Go outside and throw a rubber ball up in the air.

Does it come down? Draw what you discovered.

Up the hill
Color the picture

Draw a cat in the wagon. Is it hard to pull a cat up the hill?

Can you squeeze a toy car?

Color the picture

Circle YES or NO

How far can you shoot a small rubber band?

Color the picture

How many feet?

Can you make an electric balloon?

Color the picture

Fill up a balloon and rub it on your hair, shirt, or pants. Then see if you can get your mom's, dad's, sister's, or friend's hair to stand up!

Up the hill
Color the picture

Draw a dog in the wagon. Is it hard to pull a dog up the hill?

What goes up must come down?

Go outside and throw an egg up in the air.

Does it come down? Draw what you discovered.

Can you squeeze a virus?

Color the picture

Circle YES or NO

How far can you shoot a THICK rubber band?

Color the picture

How many feet?

Make a wave

Color the picture

Take a rope and find a friend.

Hold one end of the rope and have your friend hold the other end. Wiggle the rope.

What happens when you wiggle the rope FAST?

Draw and color a rainbow

Up the hill
Color the picture

Draw a horse in the wagon. Is it hard to pull a horse up the hill?

What goes up must come down?

Go outside and throw a feather up in the air.

Does it come down? Draw what you discovered.

Can you squeeze a soft plastic ball?

Color the picture

Circle YES or NO

Make a wave

Color the picture

Take a rope and find a friend.

Hold one end of the rope and have your friend hold the other end. Wiggle the rope.

What happens when you wiggle the rope SLOWLY?

Up the hill
Color the picture

Draw an elephant in the wagon. Is it hard
to pull an elephant up the hill?

How far can you shoot a BIG rubber band?
Color the picture

How many feet?

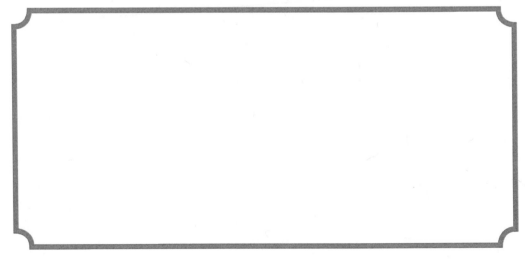

Bending light
Color the picture

Go fishing and watch the end of your fishing line. Does it bend in the water?

Looking in the Mirror
Color the picture

Can you see yourself in a mirror in a room full of light? Try it.

Circle YES or NO

Make a wave

Color the picture

Find a coiled toy and two friends. Have the friends hold each end of the coiled toy.

Squeeze the middle part of the coiled toy and let go.

What happens?

Which is faster?

a CHEETAH or a CAR

Circle one

Color the picture

What happens when you shine sunlight through a prism?

Color the picture

Make a wave

Color the picture

What do you need to do to make a SMALL wave?

Which is faster?

a CAR or a TOY CAR

Circle one

Color the picture

In the dark
Color the picture

Can you see yourself in the mirror when
the room is dark? Try it.

Circle YES or NO

Which is faster?

a SNAIL or a CAR

Circle one

Color the picture

How far can you shoot a LONG rubber band?

Color the picture

How many feet?

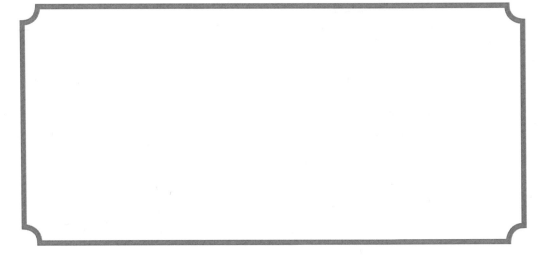

Make a wave

Color the picture

What do you need to do to make a BIG wave?

Make a wave

Color the picture

Clap your hands. What do you hear?
Did you know that sound is a wave?

Can you squeeze a hard metal ball?

Color the picture

Circle YES or NO

What sticks to a magnet?

Color the picture

Draw three things that will stick to a magnet

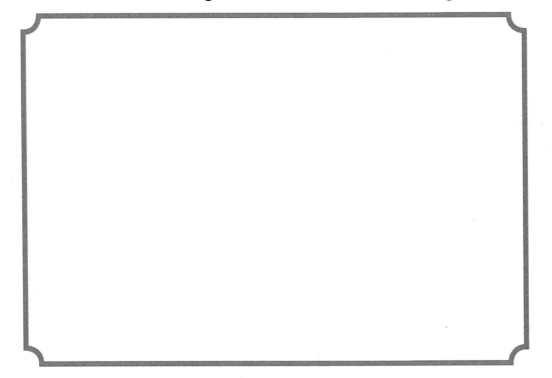

Do you live in a city?

Color the picture

Circle YES or NO

Do you live in the country?

Color the picture

Circle YES or NO

Do you live on Earth?

Color the picture

Circle YES or NO

Draw a picture of where you live

Go for a walk and find some rocks.
Draw and color them below

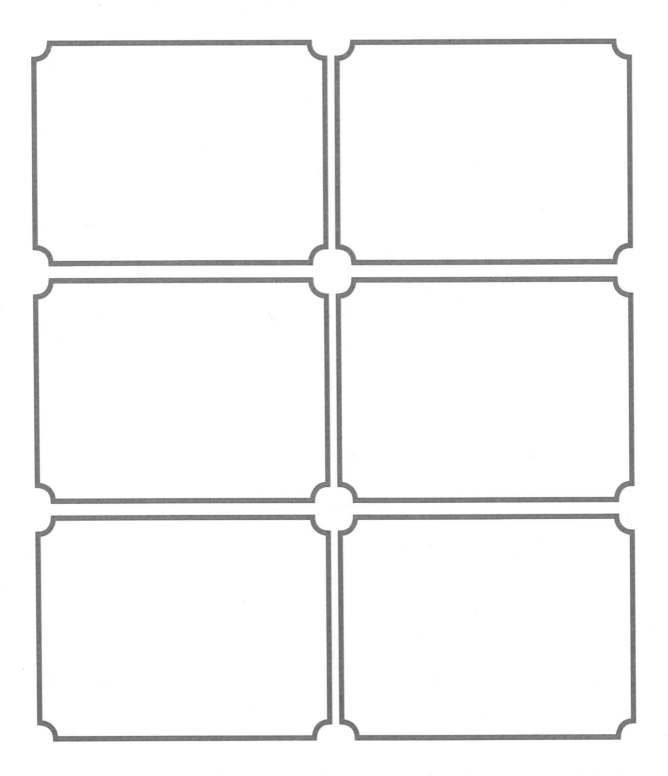

How big is Earth?

Circle the correct answer

Small enough to put
in your pocket?

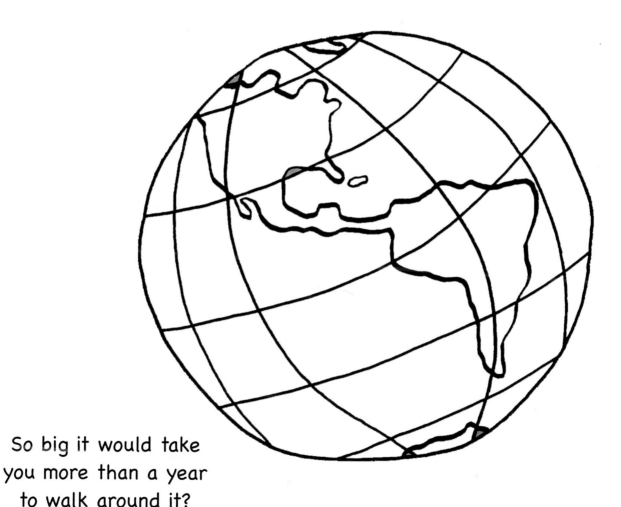

So big it would take
you more than a year
to walk around it?

Go for a walk and find some bugs.
Draw and color them below

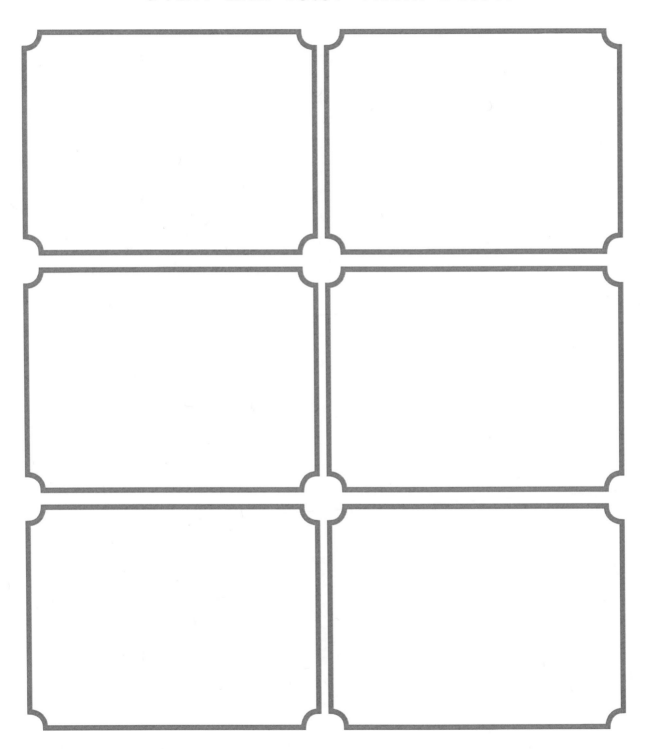

Is earth FLAT, ROUND, or SQUARE?

Circle the correct answer

What are rocks made of?

Hint: Look on page 26

Draw what rocks are made of below

When it is hot in the northern part of Earth, it is cold in the southern part of Earth.

Color the picture

Go for a walk and find some plants.
Draw and color them below

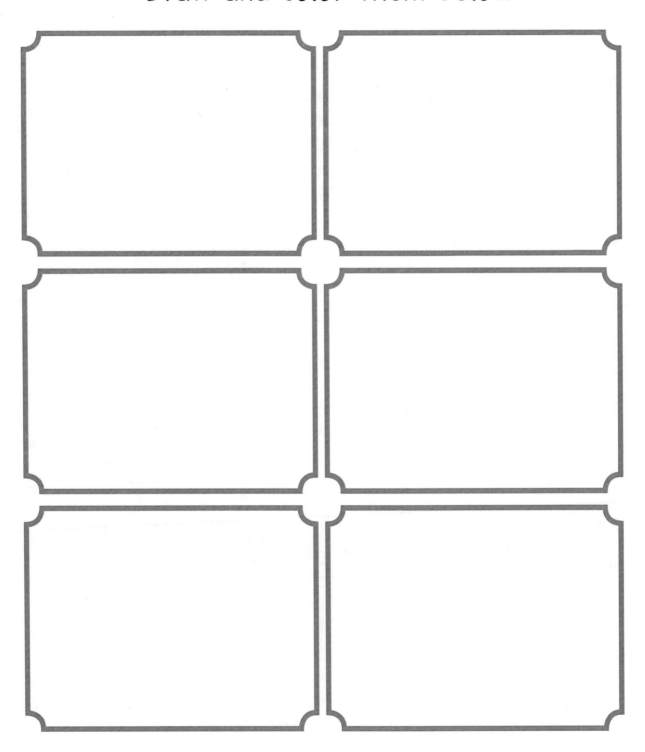

Earth spins and this gives us daytime and nightime.

Color the picture

Where does the wind come from?
Do you like the wind?

Circle YES or NO

Color the picture

Hot lava from inside the Earth makes volcanoes.

Color the picture

How long would it take you to walk around the Earth?

Color the picture and fill in the blank

Earthquakes can make big cracks in the Earth's surface.

Color the picture

Earth has layers called the CRUST, MANTLE, and CORE.

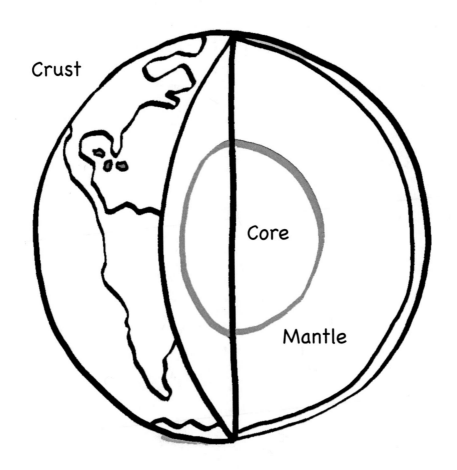

Crust

Core

Mantle

1. Color the crust blue and green
2. Color the mantle yellow
3. Color the core red

To look at rocks, geologists use special tools like a small hand lens.

Color the picture

Look at rocks with a magnifying glass or hand lens.

Draw and color what you see

Earth changes with the seasons.

Color the Spring season

Earth changes with the seasons.

Color the Winter season

Earth changes with the seasons.

Color the Summer season

Look at plants with a magnifying glass or hand lens.

Draw and color what you see

Earth changes with the seasons.

Color the Fall season

The CRUST is the outer part of Earth and is made mostly of rock.

Color the picture

How do mountains form?
Color the picture

Look at bugs with a magnifying glass or hand lens.

Draw and color what you see

What is below the ocean?

Color the picture

Draw and color animals that live in the ocean.

Draw and color animals that live in the air.

Draw and color animals that live on land.

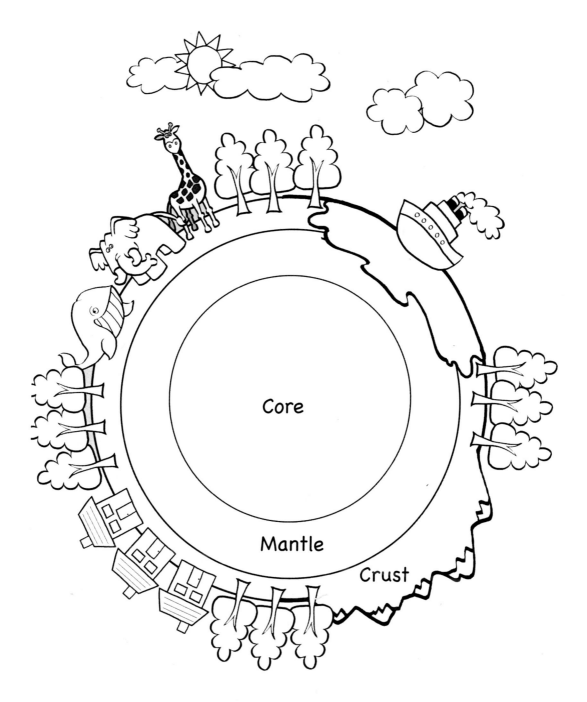

1. Color the plants and animals
2. Color the crust brown
3. Color the mantle yellow
4. Color the core red

What makes rivers?
Color the picture

What is air made of?
Color the picture

Draw Oxygen and Nitrogen atoms below

Will ships sail off the side of Earth?

Circle YES or NO

Color the picture

Why or why not?

Draw the shape of Earth

Pollution can make the air and water dirty.

Color the picture

Picking up trash helps keep the Earth clean.

Color the picture

The very top of Earth is the North Pole.

The very bottom of Earth is the South Pole.

The Earth is divided in the middle by the Equator.

North Pole

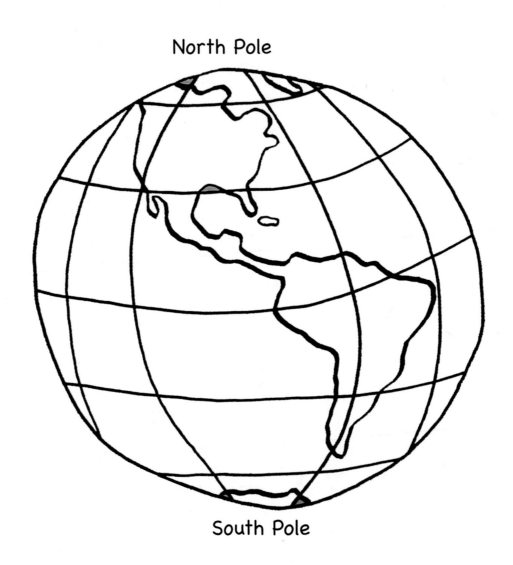

South Pole

1. Color the North Pole purple
2. Color the South Pole pink
3. Draw a line across the middle for the Equator

The needle on a compass points North.

Color the picture

ASTRONOMY

The study of SPACE

What are the lights we see in the night sky?

Color the picture

Earth is a PLANET.

Color the picture

Earth has a MOON.

Color the picture

Some people think they can see a
face on the Moon.

Go outside and look at the Moon.

Can you see a face on the Moon?

What is the MOON made of?

Color the picture

The Earth is made of rock.

Draw a picture of an Earth rock

The Moon is made of rock.

Draw a picture of a Moon rock

Draw a picture of Earth

Go outside and draw the MOON

Is it ROUND, HALF-ROUND or CRESCENT?

Draw a picture of Mars

Draw a picture of Jupiter

Draw a picture of Saturn

Draw a picture of Uranus

Go outside and draw the MOON

Is it ROUND, HALF-ROUND or CRESCENT?

Draw a picture of Neptune

Did you know?

The MOON goes around EARTH
and causes tides.

Color the picture

Did you know?

The EARTH goes around the SUN and the SUN can cause storms on EARTH.

Color the picture

Did you know?

A rocket can take people to the MOON!

Color the picture

How long would it take to drive to the SUN?

Color the picture

Go outside and draw the MOON.

Is it ROUND, HALF-ROUND or CRESCENT?

The SUN is not a planet.
The SUN is a STAR.

Draw a picture of the SUN and color it orange

There are other STARS in the sky.

1. with a white crayon, put a box around the smallest star
2. with a white crayon, put a circle around the largest star
3. color the stars

Did you know?

Lots of stars and planets together form a galaxy.

Color the picture

Did you know?

There are so many galaxies they are hard to count!

Count the galaxies in the picture

Four of the planets in our solar system are made of rock.

Draw and color the planets made of rock

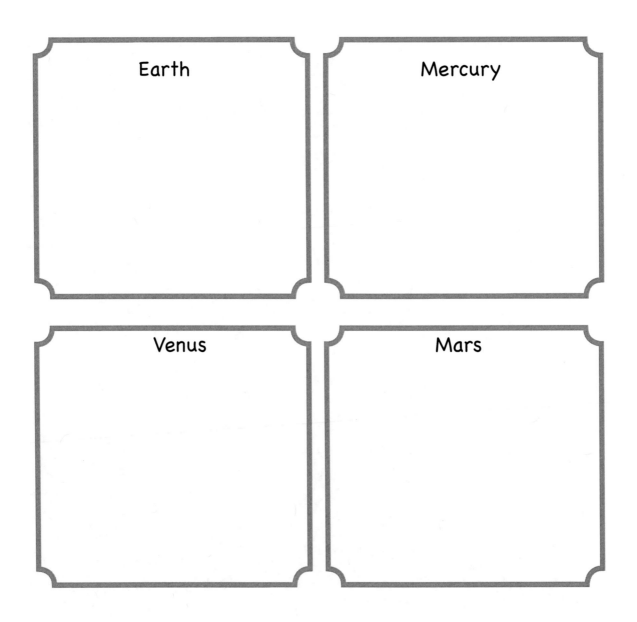

Four of the planets in our solar system are made of gas.

Draw and color the planets made of gas

Jupiter

Saturn

Neptune

Uranus

Astronomers study the stars and planets using a telescope.

Color the picture

You can see galaxies with a telescope.

Color the picture

Go outside and draw the MOON

Is it ROUND, HALF-ROUND or CRESCENT?

If you had a spaceship would you go see the Crab Nebula?

Color the picture

Did you know?

The next closest solar system is so far away we can't travel there yet.

Color the picture

Did you know?

Some stars have planets that might be like Earth.

Color the picture

Go outside and count the stars

How many did you count? Draw what you see.

Can you see any shapes made by the stars?

Draw the shapes you see and give them a name

More REAL SCIENCE-4-KIDS Books
by Rebecca W. Keller, PhD

Building Blocks Series
yearlong study program — each Student Textbook has accompanying Laboratory Notebook, Teacher's Manual, Lesson Plan, Study Notebook, Quizzes, and Graphics Package

Exploring Science Book K (Activity Book)
Exploring Science Book 1
Exploring Science Book 2
Exploring Science Book 3
Exploring Science Book 4
Exploring Science Book 5
Exploring Science Book 6
Exploring Science Book 7
Exploring Science Book 8

Focus On Series
unit study program — each title has a Student Textbook with accompanying Laboratory Notebook, Teacher's Manual, Lesson Plan, Study Notebook, Quizzes, and Graphics Package

Focus On Elementary Chemistry
Focus On Elementary Biology
Focus On Elementary Physics
Focus On Elementary Geology
Focus On Elementary Astronomy

Focus On Middle School Chemistry
Focus On Middle School Biology
Focus On Middle School Physics
Focus On Middle School Geology
Focus On Middle School Astronomy

Focus On High School Chemistry

Super Simple Science Experiments

21 Super Simple Chemistry Experiments
21 Super Simple Biology Experiments
21 Super Simple Physics Experiments
21 Super Simple Geology Experiments
21 Super Simple Astronomy Experiments
101 Super Simple Science Experiments

Note: A few titles may still be in production.

Gravitas Publications Inc.
www.gravitaspublications.com
www.realscience4kids.com